REVOLUTION
VIRTUAL TRAINING
FLAGGER'S HANDBOOK

WRITTEN AND ILLUSTRATED
BY JASON MOON

Revolution Virtual Training

Copyright © 2018 Jason Moon

All rights reserved.

ISBN-10: 1719358109
ISBN-13: 978-1719358101

CONTENTS:

Acknowledgments	1
Introduction	2
Functions	5
Equipment	7
Sign Placement/Removal	9
The Flagging Station	12
Signals	14
Operations	19
Do's & Don'ts	24
Summary	26
About Revolution Virtual Training	27
About The Author	27

ACKNOWLEDGMENTS

This handbook is based on the 2009 Edition of the Manual on Uniform Traffic Control Devices published by the U.S. Department of Transportation Federal Highway Administration.

This handbook is published as an easy to use instructional tool and field reference, however this manual is not an enforcement standard. Only the M.U.T.C.D. should be used as a standard and in the case of any discrepancy between the federal standard and this handbook, the federal standard should be strictly followed.

The text and artwork in this handbook have been developed and formatted for use with Revolution Virtual Training.

INTRODUCTION

NOUN, A person who actively controls the flow of vehicular traffic into and/or through a temporary traffic control zone using hand-signaling devices or an Automated Flagger Assistance Device (AFAD)

NOUN, A person who guards, is on lookout

SYNONYMS, caretaker, guard, sentinel, signaller

You have been chosen to be a flagger because your agency/employer feels you are physically able, mentally alert, and capable of giving directions to the motoring public.

Because you are an important part of all maintenance and construction projects, you need to know this guide well enough that it becomes second nature.

Your fellow workers and the motoring public depend upon your alertness and ability to control traffic with your STOP/SLOW paddle.

You have an important job, and it should be carried out with authority and dignity.

FLAGGER MISSION

As a flagger, you play a vital role in protecting your safety, the safety of your crew, and the safety of the motoring public from the dangers and hazards that are always present on maintenance and construction projects in public roadways.

The safety of yourself, your crew and the public are more important than any construction, maintenance or utility operation being performed.

FUNCTIONS OF A FLAGGER

Qualification to Become a Flagger:

- Ability to receive and communicate specific instructions clearly, firmly and courteously.
- Ability to move and maneuver quickly in order to avoid danger from errant vehicles.
- Ability to control signaling devices (such as paddles and flags) in order to provide clear and positive guidance to motorists approaching a work zone.
- Ability to understand and apply safe traffic control practices, sometimes in stressful or emergency situations.
- Ability to recognize dangerous traffic situations and warn workers in sufficient time to avoid injury.

Basic Functions of a Flagger:

- To protect the lives of workers
- To guide traffic safely through the work zone
- *Try* to avoid unreasonable delays to motorists
- To answer questions courteously and intelligently
- Be well informed

To assure motorist respect, your appearance is critical.

- Be clean and neat
- No Radios
- No iPods
- No Cell Phones
- Wear All PPE

FLAGGING EQUIPMENT

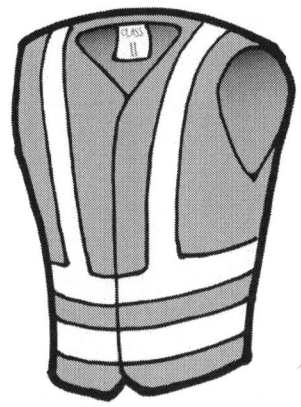

For daytime activity, you shall wear high-visibility apparel that meets Class 2 or 3 requirements. Material color shall be fluorescent orange-red, fluorescent yellow-green or a combination of the two. For nighttime activity, you shall wear high-visibility apparel that meets Class 3 requirements.

The STOP/SLOW paddle is your primary and preferred hand-signaling device. The sign shall have an octagonal shape and be a minimum 18" x 18" with letters 6" minimum height and should be mounted on a rigid handle. The paddle may be modified with high-intensity lighting per the MUTCD. At night the paddle shall be retroreflectorized.

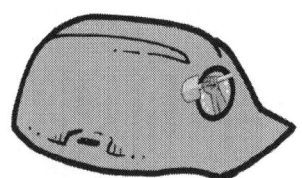

A brightly colored hat or hard hat should be considered. It will make you more visible, provides a method of communication (see Page 16) and may be required by your agency/employer.

An air horn is a good device to have in the flagger station. It is a good device to alert your co-workers of any imminent danger or errant vehicles.

In an emergency, 24" x 24" minimum red or fluorescent orange/red flags may be used until STOP/SLOW paddles can be obtained. Emergency flags should only be used in *temporary* and *unplanned* emergency situations.

The free edge of a flag should be weighted so the flag will hang vertically, even in heavy wind. Replace emergency flags immediately when STOP/SLOW paddles arrive. When used at night, emergency flags shall be retroreflectorized.

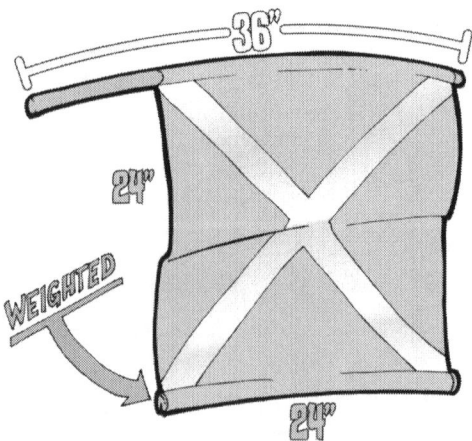

SIGN PLACEMENT-REMOVAL

Before beginning any flagging operation, advanced warning signs need to be in place. The diagram below is a typical flagging operation work zone plan.

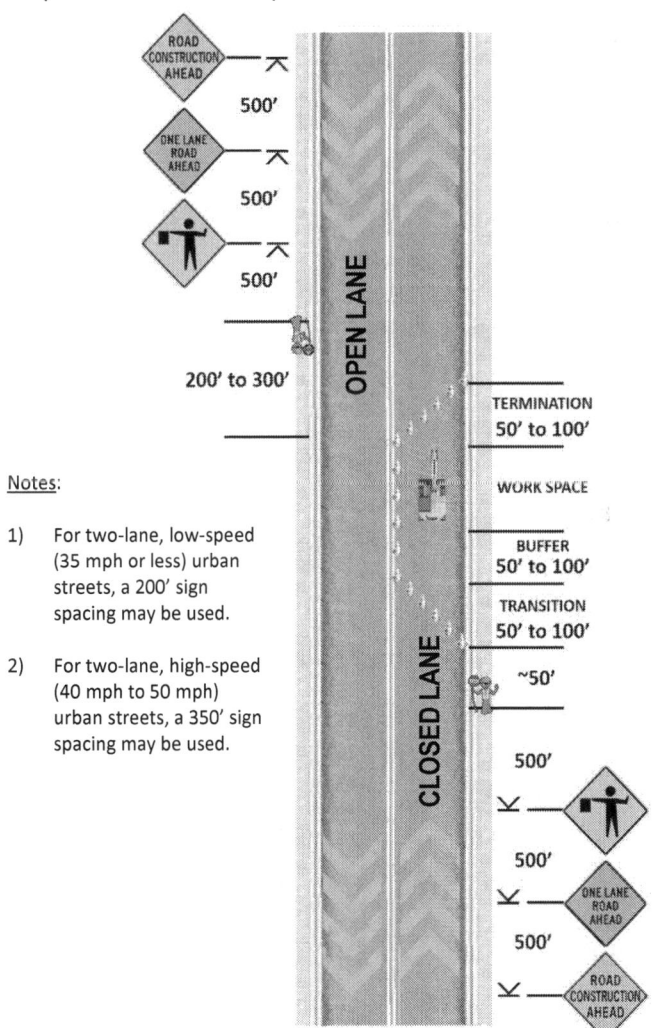

Notes:

1) For two-lane, low-speed (35 mph or less) urban streets, a 200' sign spacing may be used.

2) For two-lane, high-speed (40 mph to 50 mph) urban streets, a 350' sign spacing may be used.

WORK ZONE PLACEMENT

Work zones need to be carefully planned ahead of any work activity. The MUTCD should be consulted for correct placement and distances of devices. These TTC devices shall be in place before beginning any flagging operation.

When placing the work zone, devices should be set up from the outside working in so the devices closest to the work area are placed last as follows:
    1. Advanced Warning Area
    2. Transition Area
    3. Buffer Area
    4. Work Area
    5. Termination Area

The flagger signs should ***never*** face oncoming traffic until a flagger is in place to control traffic.

WORK ZONE REMOVAL

The flagger signs as well as any other inappropriate signs shall be removed or turned away when flagging is not being performed. This includes lunch breaks . The motoring public shall never see a flagger sign without seeing a flagger working after it.

Do Not Mislead The Public.

When removing the work zone, devices should be removed from the inside working out so the devices closest to the work area are removed first as follows:
1. Termination Area
2. Work Area
3. Buffer Area
4. Transition Area
5. Advanced Warning Area

The flagger signs should *never* face oncoming traffic after the flagger discontinues flagging operations.

THE FLAGGING STATION

You should always stand *alone in a highly visible location*. If your flagger station is too close to other workers and work vehicles, you won't be easily seen by motorists.

Your flagging location is always on the shoulder of the road. **NEVER,** never stand in the path of oncoming traffic.

DO NOT stand inside any roadside obstacle (guardrail, parked car, bridge, fence, etc.). Always know your escape route.

When relieving another flagger, quickly pass the paddle and any other equipment and leave the flagger station. Multiple people are not permitted in a flagger station even during personnel changes. Flagger's must be alone in their station.

When flagging near a vertical curve (hill), choose a flagger position in advance of the crest of the hill to maintain good sight lines with oncoming traffic.

When flagging near horizontal curve, choose a flagger position in advance of the curve to maintain good sight lines with oncoming traffic.

The MUTCD gives standard distances for device placement and flagger station placement. However, the standard also empowers staff to adjust those distances within reason to accommodate field conditions. Vertical and horizontal curvature of road surfaces is a justifiable reason to lengthen or shorten device placement distances in the interest of safety.

Communication

Communication between flaggers is difficult under compromised conditions. It is critical that flaggers have some system of communication in place before beginning a flagging operation. Two-way radios are the best way to maintain communication. If they are not available or are not functioning reliably, relay flaggers, a pilot car or flag carrying can still be used. Two-way radios should also have a back-up plan for equipment failure.

SIGNALS

Stopping Traffic

Stand safely in your flagger position on the shoulder of the road facing oncoming traffic. Never turn your back to traffic, never stand in the path of oncoming traffic. Square your shoulders to traffic to expose as much of your vest as possible to increase your visibility.

Hold the paddle away from your body with your right hand displaying the STOP side to oncoming traffic with the sign on or near the edge of the pavement. Raise your left hand, palm facing approaching vehicles, and make eye contact with the driver.

AVOID SCREECHING HALTS.

Change to the STOP only if an approaching vehicle has plenty of distance to gradually stop. When workers in the work area hear screeching halts too often, they become conditioned to not react to the noise.

Releasing Traffic

Stand safely in your flagger position on the shoulder of the road facing oncoming traffic with your paddle turned to STOP. While you continue facing traffic, alternate between eye contact with the driver and watching over your shoulder for the "All Clear" signal from the other flagger. You MUST wait for the "All Clear" signal.

Closed Lane

After you receive the "All Clear" signal it is safe to release your traffic. You will turn the paddle to the SLOW side with your right hand keeping the sign on or near the edge of the pavement. With your left hand, signal drivers to cross the center line into the available lane. Move your left hand in a sweeping motion across your body, pointing at the traffic lane drivers should travel through. Continue signaling until you are ready to safely stop traffic.

Open Lane

After you receive the "All Clear" signal it is safe to release your traffic. You will turn the paddle to the SLOW side with your right hand keeping the sign on or near the edge of the pavement. With your left hand, signal drivers to proceed in the open lane. Move your hand in a sweeping motion between your body and the paddle, pointing at the open lane. Continue signaling until you are ready to safely stop traffic

Slowing Traffic

Stand safely in your flagger position on the shoulder of the road facing oncoming traffic. With the paddle displaying SLOW, keeping your left hand away from your body with your palm facing the ground, slowly raise and lower your arm. This signal should be alternated between release signals.

"All Clear" Signal

Communication is a critical part of safely flagging traffic. With no radio communication or a radio failure, flaggers must communicate visually.

To signal "All Clear", traffic must be completely stopped and it must be clear for the other direction to be released. With traffic stopped, use your left hand to lift your hard hat to the other flagger.

Emergency Stopping Traffic

When stopping traffic, extend the flag into the roadway with your right hand while standing on the shoulder of the road. Use your standard stop signal with your left hand.

Emergency Releasing Traffic

When releasing traffic, drop the flag to your side with your right hand while standing on the shoulder of the road. Use your standard open or closed lane release signal with your left hand. Never use the flag to motion traffic through.

Emergency Slowing Traffic

To alert and slow traffic, use your right hand to wave the flag from the ground to shoulder height while standing on the shoulder of the road. Use your standard slowing signal with your left hand.

Walking Into The Road

Generally, you will stay on the shoulder of the road even after you have safely stopped the first vehicle. This is always the only safe and correct flagging position. However, if there is an oversized vehicle near the front of the traffic line so that cars near the back cannot clearly see your STOP/SLOW paddle, you may leave the shoulder temporarily if you can do so safely.

Once the first vehicle is completely stopped, keep eye contact with the driver and move toward the center line. Do not cross the center line and always keep watch for traffic approaching from behind you. While continuing to keep eye contact with the first stopped driver you will lean your paddle out past the oversized vehicle but do not put yourself in danger. Do not put any part of yourself over the center line.

When releasing traffic, you will move back to the shoulder with your paddle remaining on STOP and maintaining eye contact with the first driver. Once you are safely back on the shoulder, you may release and motion traffic to proceed through the work zone.

Never stand in the path of oncoming traffic.

FLAGGING OPERATIONS

1) **Single Flagger**
   A single flagger operation can sometimes be used to control traffic but the work area must meet specific requirements. The street shall be straight and level through the work area. Standing on the shoulder directly opposite the work area, The flagger must have clear visibility to both directions of traffic.
   The street must also be "low-volume". To determine if the street meets standards for low volume, count the number of cars that pass the intended work area in a five minute period. To be considered low-volume, the street shall have less than 3 cars in 5-minutes.
   If the street meets requirements for a single flagger operation, the flagger directs both directions of traffic with a single STOP/SLOW paddle.

2) **Two-Flaggers**
Two-flagger operations are the most common type of flagging operation. This operation uses a flagger on each end of the work zone to control traffic flow. In order to avoid confusion, one flagger should be designated as the lead flagger to expedite coordinating decisions.

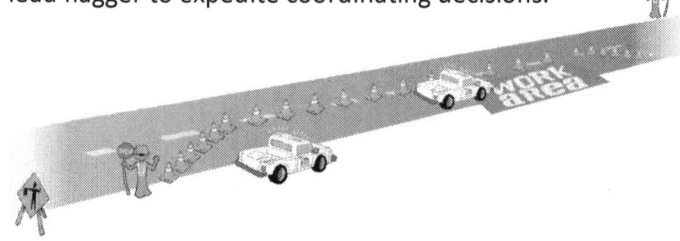

The critical part of a two-flagger operation is communication between flaggers. Communication can be maintained by one of the following:
- Visual Contact – can only be used when flaggers are close enough to read each other's STOP/SLOW paddles and see "All Clear" signals. The work zone should be free of horizontal and vertical curves.
- Two-Way Radio – Radio communication is the best way for flaggers to communicate even if visual contact is possible. However, radios require a backup plan in case of equipment failure.
- Flag Carrying – The flagger asks the driver of the last vehicle to pass a flag/baton to the other flagger to signal "All Clear".

Traffic shall only be released after the "All Clear" signal from the other flagger.

If there is ever a doubt, STOP ALL TRAFFIC!

3) **Pilot Car**
   A pilot car is used to guide a queue of vehicles through a work area or detour. A pilot car is best used if the work zone or detour is lengthy, unclear or when channelizing is ineffective.
   In a pilot car operation, a flagger holds traffic at each end of the work zone until the pilot car arrives to lead the queue of cars through the work zone.

   A safe turn-around must be available at each end of the work zone. Flaggers still need to maintain communication to identify the last vehicle in the queue. Never allow a late vehicle to catch up with the queue. Traffic at intersections and side roads should also be flagged when using a pilot car.

4) **One-Directional Control**
A single flagger can control one direction of traffic when shoulder work will temporarily block the roadway but the other direction can safely continue without being stopped. This operation is frequently used for loading/unloading materials for a work zone outside of the roadway.

The flagger should stop traffic using normal signals. Remember eye contact and no screeching stops. Release traffic once work no longer blocks the travel lane and it is safe to do so.

When releasing traffic, step back from the edge of pavement and lower the paddle to your side with the STOP side facing you to avoid confusing released traffic. This operation should be completed from the shoulder of the street. Never stand in the path of oncoming traffic.

5) **Emergency Flagging**
Emergency flagging has unique signals which can be used only until STOP/SLOW paddles arrive (See Page 18). Any emergency flagging operation shall be *temporary* and *unplanned*. When used at night, flags shall be retroreflectorized.

Replace flags with STOP/SLOW paddles as soon as they arrive.

6) **Night Flagging**

Night Flagging procedures are generally the same as daytime except for some equipment changes. Class II safety vests shall be replaced with a Class III outfit.

Retroreflectorized STOP/SLOW paddles, channelizing devices and signs should be used when flagging at night. Additional auxiliary lighting should also be considered during night operations.

The flagging signals are the same as a daytime operation except that the flagging station shall be illuminated.

THE DOS AND DONTS OF FLAGGING

DON'T Stand in the path of oncoming traffic.

DON'T Stand in the shade, over a hill or around a curve.

DON'T Stand near equipment or coworkers.

DON'T Leave your flagging position until relieved.

DON'T Have unnecessary conversations.

DON'T Bring cell phones or electronic devices.

DON'T Forget to turn inappropriate signs when done flagging.

DO Stay alert at all times.

DO Use clear standardized signals.

DO Stand on the shoulder of the road.

DO Stand ALONE.

DO Treat motorists courteously.

DO Use proper signs and equipment.

DO Know your way out.

DO Be prepared for the weather.

SUMMARY

Hopefully you have learned a lot about flagging safely. As you've seen, there is a lot to know BEFORE you begin working as a flagger. Good traffic control is an engineered solution based on uniformity, precision and standardized practices. You must always use your knowledge, wit and skills. Flagging is not easy work. It can be boring and tedious work at times. However, as a flagger, you have the responsibility of serving as the link between motorists and the work crew. Your job is to protect and safeguard hundreds or even thousands of lives as they work and travel within your work zone.

Remember these important points:
- Respect, courtesy and appearance are all critical to flagging.
- Motorists should never see a flagger sign without seeing a certified flagger after it.
- Do not stand inside obstacles, always know your way out.
- Never stand in the path of oncoming traffic.
- Avoid screeching halts.
- If you are unsure at any time, STOP ALL TRAFFIC.
- Never release traffic without the "All Clear".
- Retroreflectorized and lighted at night.

And finally, all this talk about standardization and uniformity can be overwhelming. After you flag correctly for a while, things will be less confusing. You'll develop a rhythm and flagging correctly will become second nature. Just make your best effort every day and every vehicle. We wish you luck, success and safety.

ABOUT REVOLUTION VIRTUAL TRAINING

Revolution Virtual Training is an internet-based training academy founded to bring high-quality construction industry training to municipalities, agencies and companies that don't have easy access to it by other means. R.V.T. is a training solution that works for even the smallest company or agency. Our classes are based on years of experience and lead by instructors well-rehearsed in delivering information in a way that relates well to everyone.

ABOUT THE AUTHOR

Jason Moon has worked in the construction industry as an equipment operator, engineering inspector or project manager since 1995. He began instructing procedures and safety in 2003 and has now certified hundreds of flaggers in his home state of North Carolina. He currently leads a staff of forty performing maintenance and construction on public roadways on a daily basis.

Jason's earliest jobs in the construction industry were to pay his way through college where he obtained a B.A. in Interactive Media Arts, however this handbook is the first combination of his artistic skills and the construction work and knowledge that paid for those skills.

Made in the USA
Las Vegas, NV
08 February 2025